Natacha Yacinthe

The Marine Planning Activities of United States Ports

Natacha Yacinthe

The Marine Planning Activities of United States Ports

LAP LAMBERT Academic Publishing

Impressum / Imprint

Bibliografische Information der Deutschen Nationalbibliothek: Die Deutsche Nationalbibliothek verzeichnet diese Publikation in der Deutschen Nationalbibliografie; detaillierte bibliografische Daten sind im Internet über http://dnb.d-nb.de abrufbar.

Alle in diesem Buch genannten Marken und Produktnamen unterliegen warenzeichen-, marken- oder patentrechtlichem Schutz bzw. sind Warenzeichen oder eingetragene Warenzeichen der jeweiligen Inhaber. Die Wiedergabe von Marken, Produktnamen, Gebrauchsnamen, Handelsnamen, Warenbezeichnungen u.s.w. in diesem Werk berechtigt auch ohne besondere Kennzeichnung nicht zu der Annahme, dass solche Namen im Sinne der Warenzeichen- und Markenschutzgesetzgebung als frei zu betrachten wären und daher von jedermann benutzt werden dürften.

Bibliographic information published by the Deutsche Nationalbibliothek: The Deutsche Nationalbibliothek lists this publication in the Deutsche Nationalbibliografie; detailed bibliographic data are available in the Internet at http://dnb.d-nb.de.

Any brand names and product names mentioned in this book are subject to trademark, brand or patent protection and are trademarks or registered trademarks of their respective holders. The use of brand names, product names, common names, trade names, product descriptions etc. even without a particular marking in this works is in no way to be construed to mean that such names may be regarded as unrestricted in respect of trademark and brand protection legislation and could thus be used by anyone.

Coverbild / Cover image: www.ingimage.com

Verlag / Publisher:
LAP LAMBERT Academic Publishing
ist ein Imprint der / is a trademark of
OmniScriptum GmbH & Co. KG
Heinrich-Böcking-Str. 6-8, 66121 Saarbrücken, Deutschland / Germany
Email: info@lap-publishing.com

Herstellung: siehe letzte Seite /
Printed at: see last page
ISBN: 978-3-659-62610-4

Copyright © 2014 OmniScriptum GmbH & Co. KG
Alle Rechte vorbehalten. / All rights reserved. Saarbrücken 2014

The Seaport Planning Activities of American Association of Port Authorities (AAPA)-member Seaports ("Portganisms")

The Seaport Planning Activities of American Association of Port Authorities (AAPA)-member Seaports

By

Natacha J. Yacinthe, Ph.D., AICP
Broward County Port Everglades

April 2014

The Seaport Planning Activities of American Association of Port Authorities (AAPA)-member Seaports ("Portganisms")

BIOGRAPHY

Natacha J. Yacinthe, Ph.D., AICP

Dr. Natacha J. Yacinthe is the Principal Seaport Planner and Master/Vision Plan Project Manager for Port Everglades. Dr. Yacinthe is active with the American Association of Port Authorities (AAPA) and serves on the Maritime Economic Development Committee, and Operations Committee. She is also an adjunct professor at Barry University in Miami, Florida, teaching university graduate and undergraduate courses in Research Methods, Urban Planning, Ethics, and Public Policy. Dr. Yacinthe has also taught courses at Florida Atlantic University. Dr. Yacinthe has worked with the Florida Association for Volunteer Action in the Caribbean and the Americas (FAVACA) at the request of the United States Agency for International Development (USAID) to conduct a waste management and planning assessment for the city of Jacmel, Haiti. Her work in Haiti involved conducting an assessment on waste management for that city. Her assessments and planning resulted in a 25-page report with recommendations to the Haitian government on the creation of a planning division and sustainable planning practices. Dr. Yacinthe has lived in Cape Town, South Africa, where she worked with the South African Planning Institution while conducting research for her Master's thesis. She conducted research on Input-Output model to evaluate economic benefits and impacts of certain commercial and retail uses along proposed activity streets in the former township of Langa, located in Cape Town, South Africa.

Dr. Yacinthe has presented at numerous conferences including the Transportation Research Board (TRB) in Washington, D.C., the Environmental Protection Agency (EPA), the America Planning Association, and the Southeast Conference for Public Administration (SECOPA). She is a member of the American Institute of Certified Planners (AICP) and the American Association of Planners (APA), including the Florida Chapter.

The Seaport Planning Activities of American Association of Port Authorities (AAPA)-member Seaports ("Portganisms")

Dr. Yacinthe served as the chair of the Miami-Dade Planners Technical Committee. She is currently a member of Barry University's Public Administration Advisory Board. Dr. Yacinthe's previous planning experience includes working for the cities of Miramar, Hollywood, Opa-Locka and North Miami, the city where she proposed and obtained City Commission approval to implement a planning and review fee increase that resulted in $55,000 in additional revenue to the Community Development Department within six months.

Dr. Yacinthe received her Bachelor's Degree in Public Administration and Master's Degree in Environmental and Urban Systems (MEUS) from the Florida International University College of Engineering. She was awarded a Ph.D. in Public Administration from Florida Atlantic University. She is also a three-year recipient of the Dwight D. Eisenhower Transportation Fellowship. In order to assist disadvantaged university students majoring in Urban Planning, Public Administration and Public Policy, Dr. Yacinthe has established the *Dr. Natacha J. Yacinthe Scholarship Fund*.

The Seaport Planning Activities of American Association of Port Authorities (AAPA)-member Seaports ("Portganisms")

ACKNOWLEDGMENTS

First, completion of the research paper would not be possible without the help of many people. My deepest appreciation goes to the AAPA-member seaports located in the United States and in Canada who took the time to reply to the survey questionnaire. Cesar Florian, Manuel Acosta and Professor Jim Poulos from Barry University all offered their time and assistance with the Statistical Package for Social Sciences (SPSS) and statistical assistances. I am grateful. To the 2016 PPM group and the administration at Broward County's Port Everglades for providing the time to prepare, research, survey, code, recode, analyze, and write the final paper. Thank you to the American Association of Port Authorities. My appreciation is to Broward County. I thank my "Hollywood Hills" team that provided a review when my eyes were seeing what I presumed were ships over and over again. I'm indebted. To my family for their patience and sense of humor; I'm especially thankful to my mom, Jacqueline Michel, for never asking why I needed one more hour, out of the house, in order to complete my writing while she babysat the grandkids; Drahcir, and twins Kayis and Ysatis Yacinthe. Thank you.

The Seaport Planning Activities of American Association of Port Authorities (AAPA)-member Seaports ("Portganisms")

ABSTRACT

Planning is an essential process for public organizations. Seaport planning is both complex and significant for the communities they serve. This proposed research provides a much-needed window on the characteristics of specialized seaport planning departments and the type of funding sources utilized to implement Master/Vision Planning projects. A Master Plan is a comprehensive, long-range plan intended to guide growth and development of a community or region and it can also be specific to the development of a seaport. Surveys of American Association of Port Authorities (AAPA)-membered seaports were conducted from May 25, 2013 to October 25, 2013. Survey participants consisted of the 80 AAPA member seaports Chief Executives/Directors or their designees. Often times whenever you hear of plans, one would assume that a planner is involved. The proposed research is designed to address three major questions:

- What are the characteristics of AAPA-member seaport planning departments, divisions or sections? Where are the planners within a seaport? Are planning activities located mostly in the Chief Executive/Port Director's Office?

- How are the projects identified for implementation in the adopted Master Plan funded or financed? Have all AAPA-member seaports adopted a Master/Vision Plan, and, if so, how often are these Master/Vision Plans updated?

- Is there a relationship between planning activities occurring within the seaport and which projects are eventually implemented?

The Seaport Planning Activities of American Association of Port Authorities (AAPA)-member Seaports ("Portganisms")

TABLE OF CONTENTS

	PAGE
BIOGRAPHY	2
ABSTRACT	5
INTRODUCTION	9
METHODOLOGY	14
DESCRIPTIVE ANALYSIS	16
RESULTS AND DISCUSSIONS	35
SIGNIFICANCE OF STUDY	41
REFERENCES	42
APPENDICES	43
Appendix A: The Correlation Table	43
Appendix B: The Chi-Square Table	52
Appendix C: Seaport Survey Instrument	55

The Seaport Planning Activities of American Association of Port Authorities (AAPA)-member Seaports ("Portganisms")

List of Tables

Page

Table 1.1: Gender..16

Table 1.2: Work at Port..18

Table 1.3: Type of Port..18

Table 1.4: Landlord and Operation Ports...18

Table 1.5: Acreage..19

Table 1.6: Number Employed..19

Table 1.7: Have Master plan..20

Table 1.8: Department Responsible to Update Plan..............................22

Table 1.9: Where Planning Occurs..23

Table 1.10: Planning Section of Division...24

Table 1.11: Planning Outsourced..24

Table 1.12: Primary Planning Occurs..25

Table 1.13: Adopting Master plan..26

Table 1.14: Number of Planners Employed...27

Table 1.15: Number of Planners in Planning Department.....................26

Table 1.15: Master plan Website...27

Table 1.17: Capital Improvement Plan...27

Table 1.18: Raise Capital..27

Table 1.19: Raise Capital..28

Table 1.19 Raise Capital...28

Table 1.20: Raise Capital..28

Table 1.21; Primary Source...29

Table 1.22: Primary Source...29
Table 1.23: Primary Source...29
Table 1.24: Primary Source...29
Table 1.25: Primary Source...29
Table 1.26: Ensuring Cost for MP Projects..30
Table 1.27: Affordability Analysis of Master plan Projects..31
Table 1.28: Aligned Budgets and Master plan...32
Table 1.29: In-House Planning...33
Table 1.30: Outsourcing of Planning Activities..35
Table 1.31: Stakeholder Engagement Approach...36

List of Figures

Figure 1.1   Age of Respondents..17
Figure 1.2   How Often Master Plan is Updated..21
Figure 1.3   Seaport Departments Responsible for Updating Master Plan...................23
Figure 1.4   Members Serving on Port Authority or County Boards.............................33

The Seaport Planning Activities of American Association of Port Authorities (AAPA)-member Seaports ("Portganisms")

Introduction

Ports constitute an important economic activity in coastal and suburban areas. The higher the throughput of goods, services, and passengers each year, the more infrastructure, provisions and associated services are required. Ports bring varying degrees of benefit to the local and regional economies. Ports are also important for the support of economic activities in the environs, since they act as crucial connections between sea and land transport. As a broker of jobs, ports do not only serve economic and social functions but, to some extent, political functions. Seaports are not only critical economic engines, but living "organisms". *In biology, an organism is defined as a contiguous living system of which, in at least some form, capable of response to stimuli, reproduction, growth and development, and maintenance as a stable whole.* One of the basic parameters of an organism is its life span. Some organisms live as short as one day, while some can live thousands of years. In order for a seaport to continue to thrive or develop, it takes constant care and understanding and commitment from the employees, shippers, port tenants, residents, elected and public officials, and other stakeholders to address the fundamentals of these systems. Ports, in essence, are what I've coined living *"portganisms,"* and they require attention, security, care and protection in order to facilitate their development, growth, maintenance and sustainability. In the past several years, specialized planning and adoption of Master/Vision Plans are significant activities of United States seaports. This paper seeks to identify the planning and planning-related activities of seaports, if the seaports in the United States develop and adopt Master/Vision Plans. For the purpose of this research, a Master/Vision Plan is an evolving, long-term planning document usually reflecting a clear vision created and adopted in an open process. The Master/Vision Plans usually consist of six elements including:

- Element 1: Existing Conditions
- Element 2: Market Assessment
- Element 3: Facility Assessment/Plan Development
- Element 4: Strategy Development
- Element 5: The Final Plan

The Seaport Planning Activities of American Association of Port Authorities (AAPA)-member Seaports ("Portganisms")

- Element 6: Plan Implementation

The research intends to review and discuss the characteristics of seaport planning departments. The goal of the paper is to provide much-needed information into the planning activities of AAPA-member seaports; and to encourage seaports to develop Master/Vision plans.

Daniel Burnham, considered the father of urban planning, famously quoted, *"Make no little plans; they have no magic to stir men's blood and probably themselves will not be realized. Make big plans; aim high in hope and work remembering that a noble, logical diagram, once recorded, will not die."* Planning is a critical component and activity. A plan can take a variety of forms including strategic plans, comprehensive plans, neighborhood plans, economic development, historic preservation or Master/Vision plans. Master/Vision Plans that guide policy decisions and seaport planners are often also responsible for enforcing the chosen policies in the plan. Most seaports' planning activities are identified in the development and adoption of their Master/Vision plans. Planning provides direction on proposed development and a seaport's Master/Vision plan serves as a guide to proposed and anticipated development projects, market conditions and characteristics. This proposed study is designed to address research on the planning activities of the 80 AAPA-member seaports located throughout the United States. Recent overviews of the literature and discussions with participants at PPM conferences suggest that there is currently minimal research on the characteristics of seaport planning department is and the seaports' adopted Master/Vision Plans.

According to AAPA, almost all of the top 80 U.S. public seaport agencies are members of, and are represented by, AAPA. These seaports established by enactments of state government and/or public port agencies develop, manage and promote the flow of waterborne commerce and act as catalysts for economic growth. They develop and maintain the terminal facilities for intermodal transfer of cargo between ships, barges, trucks and railroads. Port authorities also lease land, and in some cases build and maintain facilities for the growing cruise, excursion, and ferry passenger industry.

The Seaport Planning Activities of American Association of Port Authorities (AAPA)-member Seaports ("Portganisms")

In addition to maritime functions, port authority activities may also include airports, bridges, tunnels, commuter rail systems, inland river or shallow draft barge terminals, industrial parks, foreign trade zones, world trade centers, terminal or short-line railroads, shipyards, dredging, marinas and other public recreational facilities. In addition, a few ports offer ferry services, such as Port of Portland (Maine), Delaware River Port Authority and Port of San Francisco. Many ports, including Port Miami and Port Everglades, are considering adding ferry operations if the market can sustain such operations.

According to a survey conducted by AAPA in the United States, public ports work closely with private industry, in the development and financing of marine terminals and other maritime-related facilities. Seaports also play a critical role in our national security, peace-keeping and humanitarian efforts around the world. Public ports also serve as sponsors of federal navigation projects that benefit all maritime interests. During the first five months of 2012, the 63 U.S. ports (out of 80 surveyed) that returned completed AAPA Port Infrastructure Investment questionnaires reported that they, and their private sector business partners, planned to invest an estimated $46 billion for infrastructure during the five-year period between 2012 and 2016 (**AAPA Summary U.S. Public Port Survey 2012-2016**).

Purpose of the Study

The intent of this study is to learn about characteristics of planning activities and departments located within AAPA-member seaports. The focus is on the types of planning activities of seaports, as well as on the types of funding sources seaports utilize in order to implement projects identified in the 5-year capital improvement sections of their Master/Vision plans. The research also divulges where planning activities are conducted within a port's organization, and if there is a significant relationship among seaport planning and various other variables including, number of Board members, funding sources and size of seaports.

The Seaport Planning Activities of American Association of Port Authorities (AAPA)-member Seaports ("Portganisms")

Nature and Limitations of the Study

According to AAPA, there are 360 commercial seaports located in the United States and 18 seaport authorities that are located in Canada. Due to timing and funding limitations, the researcher conducted a survey of Chief Executives at AAPA-member seaports in Canada and the United States. Currently, there are 80 seaports in the United States and 10 seaport authorities in Canada who are members of the AAPA. A total of 80 surveys were mailed and 42 seaports responded to the survey which resulted in a 54% response rate. The results from the survey consisted of a representative sample of AAPA seaports. Based on a May 2013 seaport list provided by the AAPA, there are 15 AAPA-member Canadian ports, 11 AAPA-member Caribbean seaports, and 33 AAPA South American seaport members. Due to the time and financial limitations, the Caribbean and South American seaports were not part of the study but could potential be surveyed in the future.

Contribution to Knowledge

Planning is vital to all public organizations. Planning can occur in all areas of an organization. Within seaports where are the planning functions conducted? Is it within Economic Development Departments, Environmental Management Departments, in the Chief Executive's Office, Public Works Department or outsourced to private consulting firms? Second, are the adoptions of Master/Vision Plans the primary function of the planning staff? As a result of this research in conjunction with my time abroad, my intent is to gather and organize information that will be useful for future researchers. There is an opportunity to share the results of this research project with PPM individuals and a broader audience in the hope that such findings will benefit the seaports who desire to learn more about seaport planning functions.

The Seaport Planning Activities of American Association of Port Authorities (AAPA)-member Seaports ("Portganisms")

Research Questions

The proposed research is designed to address three major questions:

- What are the characteristics of AAPA-member seaport planning departments, divisions or sections? Where are the planners within a seaport? Are planning activities located mostly in the Chief Executive/Port Director's office?

- How are the projects identified for implementation in the adopted Master Plan funded or financed? Have all AAPA-member seaports adopted a Master/Vision Plan, and if so, how often are these Master /Vision Plans updated?

- Is there a relationship between planning activities occurring within the seaport and which projects are eventually implemented?

Assumptions

The assumption of this research is that while seaport planning is a significant component of a port authority and other local or state government agencies, there is limited information on the activities of seaport planning departments and the roles planning plays in seaports. The research assumes that adoption and updating of Master Plans can help to guide future policy decisions and assist Chief Executives/Port Directors to prioritize and make wise use of resources.

The Seaport Planning Activities of American Association of Port Authorities (AAPA)-member Seaports ("Portganisms")

METHODOLOGY

Population and Sample

According to the United States Coast Guard, the United States is served by 360 commercial ports that provide approximately 3,200 cargo and passenger handling facilities. Each seaport is unique, depending on its location and the specific services. The type of activities at a seaport can vary greatly. For the purposes of this research, the population (N = 80) consisted of 80 Chief Executives/Port Directors of the seaports (located in the United States and Canada) that are members of the AAPA. Participants were selected from the AAPA member list of seaports provided by AAPA Headquarters in May 2013. The paper also aims to survey 81 Executive Directors at each of the AAPA-member seaports to determine if the planning activities are guided by a currently adopted Master/Vision Plan. The research sought to also identify the characteristics of an AAPA-member seaport planning department and if the primary function of the planning department is the adoption of the Master/Vision Plan. The research anticipates surveying at least the entire AAPA United States and Canada membership list. The survey strived to achieve a response rate of 50%. Achieving this response rate will mean that the survey results are representative of the entire AAPA-member seaport population. There was no guarantee that a 50% response rate and accuracy goals will be achieved since response is based on the interest level of survey recipients. However, all Port Directors and their designees were assured of privacy and confidentiality, with the option to view written results, reported in the aggregate, upon completion of the study. A total of 42 survey questionnaires were returned which resulted in a 52% response rate.

Study Population

Seaports located in the United States and Canada and that are members of the American Association of Port Authorities (AAPA).

The Seaport Planning Activities of American Association of Port Authorities (AAPA)-member Seaports ("Portganisms")

Instrumentation/Data Collections

A 30-question survey (Exhibit A) was administered by e-mail to AAPA-member seaports Executives and Port Directors from May 21, 2013 to October 25, 2013. A reminder to survey recipients was sent three weeks and subsequently one month after initial receipt. The research schedule and timetable was as follows:

Schedule/Timetable

April 2013 to May 2013	Prepared Abstract and Proposal for AAPA Approval
April 2013 to May 2013	Prepared and Finalized survey instrument
May 2013 to October 2013	Administered Questionnaire to AAPA-member Ports
October 2013 to January 2014	Input & Coded Data and Analyzed Results
October 2013 to March 2014	Prepared Final Report

Ongoing data analysis took place throughout the study. Connections between categories and trends were used to further the understanding of the characteristics of seaport planning departments. Preliminary analyses were performed and helped to determine whether there are any systematic differences. Results are determined by utilizing and comparing descriptive statistical results. The researcher also conducted correlational analysis and Chi-square tests utilizing the Statistical Packet for Social Science (SPSS) to analyze the results. However, inferential analysis was also conducted to determine if there was evidence of relationships between the variables.

Funding

Funding for the research was based on time allocated by Broward County Port Everglades to conduct the research. It was anticipated that a year would be needed to complete the study including preparation of survey instrument, pre-test, conducting the survey, inputting the data, analyzing the results and preparing the final report. The surveys were conducted from May 2013 to October 2013.

The Seaport Planning Activities of American Association of Port Authorities (AAPA)-member Seaports ("Portganisms")

DESCRIPTIVE ANALYSIS

This section provides a descriptive analysis of the results of the survey. The results are presented according to the responses received for the survey which was conducted from May 21, 2013 to October 25, 2013. Eighty AAPA-member seaports were surveyed and 42 responded, for a 52% response rate.

Survey Responders, Type of Port and Port Staff

Seventy-eight percent (78%) of survey responders were males and 19.5% were females. Another question asked about the age of the survey respondents. 7.65% of the respondent were between the ages of 26 to 35 years old, 23.08% were 36 to 45 years old, 25.64% were between 46 to 55 years old, 33.33% between 56 to 65 years old and 10.26% were 65 years old or older. Table 1 and Figure 1.1 below display the frequency and percentage.

Table 1
What is your gender?

	Number	Percent
Female	8	19.5
Male	32	78.0
No Response Given	2	0
Total	42	100.0%

The Seaport Planning Activities of American Association of Port Authorities (AAPA)-member Seaports ("Portganisms")

Figure 1.1 Age of Respondents

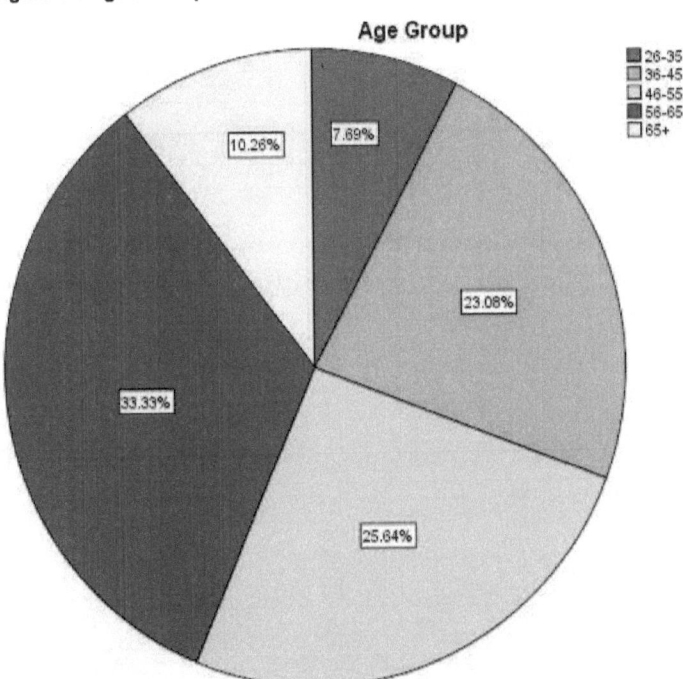

When asked how long the respondents have worked for their respective seaports, almost 5% (4.9%) reported that they worked for the port less than two years, almost 71% (31.7% and 39.0%) have worked at the port between 2 to 15 years, and 24.4% report working for the port more than 15 years. See Table 1.3 below.

The Seaport Planning Activities of American Association of Port Authorities (AAPA)-member Seaports ("Portganisms")

Table 1.2
How long have you worked for the Port?

	Number	Percentage
Less than 2 years	2	4.9
2 to 7 years	13	31.7
8 to 15 years	16	39.0
More than 15 years	10	24.4
No Response Given	1	0
Total	42	100.00%

Respondents were asked about the types of ports. More than ninety-seven percent (97.62%) worked at seaports and 2.38% (1) reported working at an in-land port (see Table 1.3 below).

Table 1.3
Type of Port

Seaport	41	97.62
Inland Port	1	2.38
No Response Given	0	0
Total	42	100.00%

More than 67% of respondents identified their seaports as both a landlord and operational seaport (see Table 1.4 below).

Table 1.4 Both Landlord and Operational Port

	Number	Percentage
Yes	13	32.5
No	27	67.5
No Response Given	2	0
Total	42	100.00%

The Seaport Planning Activities of American Association of Port Authorities (AAPA)-member Seaports ("Portganisms")

As displayed in Table 1.5, 46.3% of the seaports were less than 1,000 acres, 29.3% were 1,001 to 5,000 acres of port jurisdiction, 14.6% were between 5,001 to 10,000 acres, and 9.8% of the responding ports were larger than 10,000 acres.

Table 1.5
Total acreage of your seaport under the Port Authority's Jurisdiction

	Number	Percentage
Less than 1000	19	46.3
1001 to 5000	12	29.3
5001 to 10,000	6	14.6
More than 10,000	4	9.8
No Response	1	0
Total	42	100.00%

Respondents were asked to provide the total number of employees employed at the respective ports (see Table 1.6 below). 19.5% of ports employed fewer than 25 employees. Over 34.1% employ 25 to 99 (33.3%) employees and almost 37% (36.6%) employ 101 to 500 employees, 9.8% employ over 500 employees. While 2.4% did not respond to the question.

Table 1.6
Number Employed by the seaport?

	Number	Percentage
Approximately and Estimated		
Less than 25	8	19.5
25 to 99	14	34.1
100 to 500	15	36.6
More than 500	4	9.8
No Response Given	1	0
Total	42	100.00%

# The Seaport Planning Activities of American Association of Port Authorities (AAPA)-member Seaports ("Portganisms")

Two questions concerned the Master/Vision Plans updates and adoption. The results reveal that 78% of seaports have an adopted a Master Plan and 22% reported they do not have a Master Plan. The seaports that reported having an adopted plan were asked how often they updated their Master/Vision plans. Figure 1.2 below displays the results. About 17% of responding seaports do not update the port's Master plan, 9.76% update their master plan every two years, over 21% update every 2 to 5 years. More than 40% (43.90%) update the seaport's Master Plan every 5 to 10 years, and 7.32% report updating the Master plan every 15 years or more (See Tables 1.7, pg. 21 and Figure 1.2 below).

Table 1.7
Does your Port have an adopted Master Plan?

	Number	Percentage
Yes	32	78.0
No	9	22.0
No Response Given	1	0
Total	42	100.0%

The Seaport Planning Activities of American Association of Port Authorities (AAPA)-member Seaports ("Portganisms")

Figure 1.2
How often does your seaport update the Master plan?

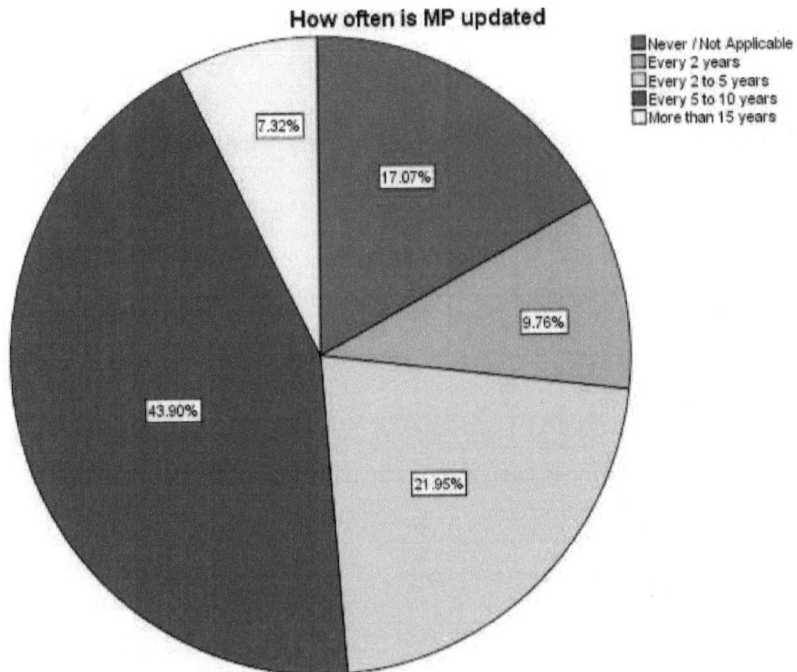

The survey asked which department within the seaport was responsible for updating the seaport's Master/Vision plan. 56.8% reported that Chief Executive/Port Director's Offices were responsible for updating the plan; 5.4% stated the seaport's Environmental or Office of Sustainability, Engineering Department or Economic Development Department was responsible for updating the plan; 2.7% reported the oversight for the master plan was conducted by the Port Authority Commissioner's Office, Operations Department, Trade Development Office, or the Port's Restoration Department. Almost 14% of the respondents identified said that the seaport's Planning

The Seaport Planning Activities of American Association of Port Authorities (AAPA)-member Seaports ("Portganisms")

Department was responsible for managing and updating the Port's Master Plan. Table 1.8 below displays the results. The results were combined, in Figure 1.3 below, noting 56.8% of Master plan adoption and updates were conducted in the Chief Executive's/Port Director's offices and 43.2% of "other" departments were responsible for managing the seaport's Master Plan.

Table 1.8
Which department is responsible for managing and updating the Port's Master Plan?

	Number	Percentage
Port Director's / Chief Executive Office	21	56.8
Environmental/Office of Sustainability	2	5.4
Port Authority / Port Commissioner's Office	1	2.7
Engineering Department	2	5.4
Operations Department	1	2.7
Planning Department	5	13.5
Economic Development	2	5.4
Trade Development	1	2.7
Chief Operating Officer/COO	1	2.7
Port Restoration Department	1	2.7
No Response Given	5	0
Total	42	100.0%

The Seaport Planning Activities of American Association of Port Authorities (AAPA)-member Seaports ("Portganisms")

Figure 1.3

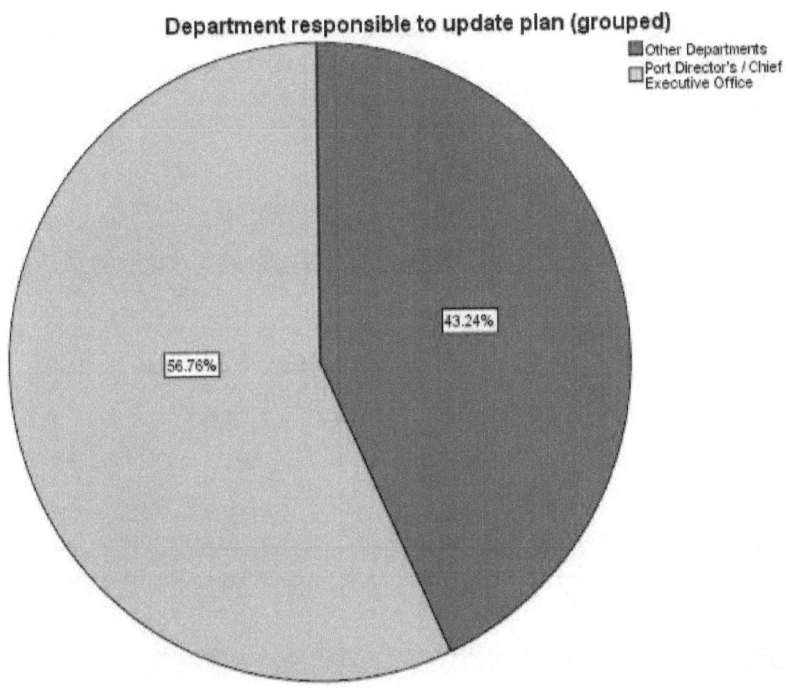

When asked where the seaport planning activities primarily occurred, sixty-eight percent (68%) of respondents revealed that the seaport planning activities were not conducted in the Planning Department, while 31.7% stated that they were (See Table 1.9 below).

Table 1.9
Are your seaports planning activities conducted under a Planning Department?

	Number	Percentage
No	28	68.3
Yes	13	31.7
No Response Given	1	0
	42	100.0%

The Seaport Planning Activities of American Association of Port Authorities (AAPA)-member Seaports ("Portganisms")

As displayed in Table 1.10, the seaports that did not have a Planning Department were asked if they had a planning section or planning division at their respective ports. Almost 64% (63.4) reported not having a Planning Division or Planning Section while 36.6% have a either a seaport Planning section or Planning division.

Table 1.10
Does your Port have a planning section or a planning division?

Planning Section or Division	Number	Percent
Yes	26	63.4
No	15	36.6
No Response Given	1	0
Total	42	100.0%

When asked if the seaport's planning activities were mostly outsourced to consulting firms, 63.4% replied no and almost 37% stated their planning activities are outsourced to a professional or consulting firm. See Table 1.11.

Table 1.11
Is your seaport planning activities mostly outsourced to a professional planning and consulting firm?

	Number	Percent
No	26	63.4
Yes	15	36.6
No Response given	1	0
Total	42	100.0%

Another question asked where specifically planning took place within a seaport. What are the primary departments where planning activities within AAPA-member ports were conducted? Fifty-five percent (55%) of seaport planning activities are conducted within the Port Director/Chief Executive Offices; almost eighteen percent (17.5%) of seaports had no official Planning Department, 10% of planning is conducted under an Engineering Department, 7.5% are conducted under an Environmental/Sustainability Department, with 5% of planning occurring in Planning Department and 5% by Port Authority's Commissioner's staff members (See Table 1.12 on pg., 26).

The Seaport Planning Activities of American Association of Port Authorities (AAPA)-member Seaports ("Portganisms")

Table 1.12
What is the primary department where *planning activities* are conducted?

	Number	Percent
Port Director's Office/Chief Executive Office	22	55.0
Not Applicable/No Official Planning Occurring at Seaport	7	17.5
Engineering Department	4	10.0
Environmental Department/Sustainability Department	3	7.5
Planning Department	2	5.0
Port Authority/Port Commission Office/Staff	2	5.0
No Response Given	2	0
Total	42	100.0%

When asked about the planning staff primary responsibilities, over 20% (21.6%) of respondents stated that the management and adoption of the Master plan was the primary function of the planning staff and nearly 78% reported adoption and management of the plan was not the primary function of the seaport's planning staff.

Table 1.13
Is managing and adopting the Master Plan the primary function of the seaport planning staff?

	Number	Percent
No	29	78.4
Yes	8	21.6
No Response Given	5	0
Total	42	100.0%

Another question was regarding the number of staff members whose job title is officially "Planner." As revealed in Table 1.14 below, 36.6% of the responding seaports had no planners working at the port, and 51.2% had between one to five planners working at the port. Almost 10% (9.8) had 6 to 10 planners and 1 port employed 11 to 15 planners.

The Seaport Planning Activities of American Association of Port Authorities (AAPA)-member Seaports ("Portganisms")

Table 1.14
How many planners are employed in the Port?

	Number	Percent
None	15	36.6
1 to 5	21	51.2
6 to 10	4	9.8
11 to 15	1	2.4
No Response Given	1	0
Total	42	100.0%

Of those seaports that employed planners, a follow-up question was how many of those planners worked within the seaport's Planning Department or Planning Divisions (See Table 1.15 below). Almost 44% stated having no seaport planners working within a Planning Department or Division. Respectively, almost 47% (46.3%) and 9.8% of seaports report employing 1 to 5 and 6 to 10 planners.

Table 1.15
How many seaport planners are employed specifically within your Port's Planning Department or Planning Division?

	Number	Percent
None	18	43.9
1 to 5	19	46.3
6 to 10	4	9.8
No Response Given	1	0
Total	42	100.0%

Table 1.16
Is there a web page/website dedicated solely to your Port's Master/Vision Plan?

	Number	Percentage
No	27	65.9
Yes	14	34.1
No Response Given	1	0
	42	100.0%

The Seaport Planning Activities of American Association of Port Authorities (AAPA)-member Seaports ("Portganisms")

SEAPORT FUNDING SOURCES AND GOVERNANCE

When asked about the primary source funding for projects identified for implementation or construction in the seaport's Master plan, almost 25% reported federal funding as the primary source, 16% reported regional or state funding as the primary source, 35% reported local government as primary source. Almost 30% stated private sector participation or funding as the primary source for project implementation. Almost 14% reported funds from the provincial province of Canada and 39% having port revenue as the primary source of funding Master plan projects.

Many seaports prepare Capital Improvement Plans (CIP), one question asked ports to provide information as to whether their Master plans have a CIP component or a 5-Year projects component. Based on Table 1.17 below, almost 80% (79.5%) of ports have a CIP or 5-Year project component within their Master Plans, while 20.5% stated they do not have either.

Table 1.17
Does your seaport Master plan have a Capital Improvement Plan (CIP) or a 5-Year project component?

	Number	Percentage
Yes	31	79.5
No	8	20.5
No Response Given	3	0
Total	42	100.0%

In addition, respondents were asked how capital or financing was raised for construction of capital projects at the seaport. This question was an open-ended question which provided respondents the ability to list the CIP funding sources for their respective seaports. Tables 1.18, 1.19, and 1.20 (below) capture the source-only results. Table 1.19 identifies the seaports that finances capital projects through the use of private sector donations or through Public Private Partnerships (P3) approach at 7.1% of seaports. In Table 1.19, almost 17% of AAPA-member ports reported federal and/or state funding as a means of financing their CIP projects. Thirty-one percent reported utilizing only the seaport's revenue or floating bonds to finance their CIP projects (See Table 1.20 below).

Table 1.18
Raise capital for funding of capital projects only by private donations

	Number	Percentage
No	39	92.9
Yes	3	7.1
No Response Given	0	0
Total	42	100.0%

Table 1.19
Raise capital for funding of capital projects only by federal/state grants

	Number	Percentage
No	35	83.3
Yes	7	16.7
No Response Given	0	0
Total	42	100.00

Table 1.20
Raise Capital for funding of capital projects only by port revenue

	Number	Percentage
No	29	69.0
Yes	13	31.0
No Response Given	0	0
Total	42	100.00

Seaports were asked to identify the primary funding sources utilized to finance projects identified in their Master Plans. As to federal funds, 7.3% seaports reported federal funding as the primary funding source, 26.8% identified state funds as the prime source for Master plan projects, almost 14.6% stated local or county funds as primary sources, and nearly 30% reported primary funds through a Public Private Partnerships (P3) and 9.8% through port revenue only.

The Seaport Planning Activities of American Association of Port Authorities (AAPA)-member Seaports ("Portganisms")

Table 1.21
What is the primary source of funding for projects identified within your Port's Master plan? Primary Source: Federal

	Number	Percentage
No	38	92.7
Yes	3	7.3
No Response Given	1	0
Total	42	100.0

Table 1.22
What is the primary source identified: *State Funds*

	Number	Percentage
No	30	73.2
Yes	11	26.8
No Response Given	1	0
Total	42	100.0

Table 1.23
What is the primary source identified: Local/County Funds

	Number	Percentage
No	35	85.4
Yes	6	14.6
No Response Given	1	0
Total	42	100.0

Table 1.24
What is the primary source identified: Public/Private Partnerships (P3)

	Number	Percentage
No	29	70.7
Yes	12	29.3
No Response Given	1	0
Total	42	100.0

The Seaport Planning Activities of American Association of Port Authorities (AAPA)-member Seaports ("Portganisms")

Table 1.25
What is the primary source identified: Port Revenue Only

	Number	Percentage
No	37	90.2
Yes	4	9.8
No Response Given	1	0
Total	42	100.0%

Tables 1.26
Who bears the ensuing costs for projects that are implemented as part of the Master plan?

	Number	Percentage
Federal		
No		75.68
Yes		24.3
No Response Given		0
Regional/State		
No	31	83.8
Yes	6	16.2
No Response given	5	0
Total	42	100.0%
Local/Municipal Government		
No	24	64.9
Yes	13	35.1
No Response Given	5	0
Total	42	100.0%
International Sector		
No	37	88.1
Yes	0	0.0
No Response Given	5	0
Total	42	100.0
Private Sector		
No	27	71.1
Yes	11	28.9
No Response Given	4	0
Total	42	100.0%

The Seaport Planning Activities of American Association of Port Authorities (AAPA)-member Seaports ("Portganisms")

Provincial Government (Canada Seaports)		
No	32	86.5
Yes	5	13.5
No Response Given	5	0
Total	42	100.0%
Port Revenues (Only)		
Yes	22	61.1
No	14	38.9
No Response Given	6	0
Total	42	100.0%

To determine the potential mechanism to finance the projected Master Plan projects, a planning and financial "affordability" analysis may be conducted by a seaport. Displayed in Table 1.27, are the results of the Affordability Analysis question. Almost 63% of seaports conduct an affordability analysis and 37.5% stated they do not conduct or do not have a port Master plan.

Table 1.27
Is there an "Affordability Analysis/Decision-Matrix" conducted when prioritizing which Master Plan projects to implement?

	Number	Percentage
No	15	37.5
Yes	25	62.5
No Response Given	2	0
Total	42	100.0%

Table 1.28
Do you align the financing of the Master plan projects with your current budget cycle or as part of the annual Operating budgets?

	Number	Percent
Yes	31	81.6
No	7	18.4
No Response Given	4	0
Total	42	100.0%

Table 1.29
For 2010-2013 fiscal years, what percentage of the Board approved Master Plan related projects were primarily implemented or prepared by in-house staff?

	Number	Percent
0% to 25%	14	37.8
26% to 50%	8	21.6
51% to 75%	4	10.8
76% to 100%	11	29.7
No Response Given	5	0
Total	42	100.0%

Table 1.30
For 2010-2013 fiscal years, what percentage of Board approved Master-plan related approved projects were primarily implemented or prepared by a consultant firm or outsourced?

	Number	Percent
0% to 25%	18	47.4
26% to 50%	5	13.2
51% to 75%	6	15.8
76% to 100%	9	21.4
No Response Given	4	0
Total	42	100.0%

The respondents were asked on the number of Port Commissioners serving on the Board: 34.1% have between 1 to 5 board members, over 56% (56.1) have 6 to 10 Board members, and almost 10% have between 11 to 15 board members serving on the board (See Figure 1.4).

The Seaport Planning Activities of American Association of Port Authorities (AAPA)-member Seaports ("Portganisms")

Figure1.4

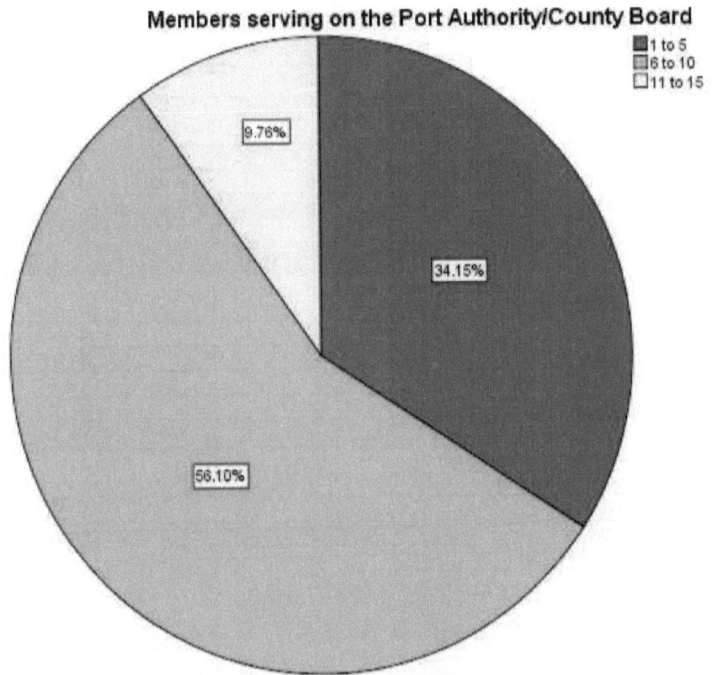

Planning involves engaging stakeholders. Respondents were asked about the methodology to engage stakeholders in the planning process including the Master plan process. As displayed in Table 1.31 (below), 38.1% stakeholder involvement is through formal meetings (public meetings, Commission meetings) with stakeholders. Almost fifty-two percent reported they engaged the stakeholders through outreach events by holding community forums. Almost seventy percent (69.2%) stakeholder engagement involved only consultation with seaport staff while 46.2% was through oral participation or contacting port staff, and 38.5% of seaports involved their stakeholders in the planning process through e-mail, web-based communication or phone surveys.

The Seaport Planning Activities of American Association of Port Authorities (AAPA)-member Seaports ("Portganisms")

Table 1.31
What is the methodology used to engage stakeholder outreach and involvement in the Master planning process?

	Number	Percentage
Stakeholder Involvement- *Formal Meetings*		
No	23	54.8
Yes	16	38.1
No Response Given	3	0
Total	42	100.0%
Stakeholder Involvement- *Outreach Events/ Forums*		
No	19	48.7
Yes	20	51.3
No Response Given	3	0
Total	42	100.0%
Stakeholder Involvement- *Consultation with Port Staff Only*		
No	12	30.8
Yes	27	69.2
No Response Given	3	0
Total	42	100.0%
Stakeholder Involvement- *Oral Participation*		
No	21	53.8
Yes	18	46.2
No Response Given	3	0
Total	42	100.0%
Stakeholder Involvement- *Written Participation Only (Surveys/E-mail Feedback, etc.)*		
No	24	61.5
Yes	15	38.5
No Response Given	3	0
Total	42	100.0%

The Seaport Planning Activities of American Association of Port Authorities (AAPA)-member Seaports ("Portganisms")

RESULTS AND DISCUSSION

This section present the results of the analysis conducted to test and analyze the variables in the questionnaire. Inferential statistics essentially was conducted in order to determine generally what we observed and relate it to the general AAPA-member seaport population or seaport survey respondents which in this case, were the remaining seaports that did not or were unable to complete the survey. Two inferential analyses were conducted. Correlation and Chi-Square were analyzed in order to determine the correlation (significance and relationship) and the Chi-Square (association/effect) for each variable from the questionnaire. Correlation analysis measures a relationship or association between variables; it does not define the explanation or its basis. Correlation analysis is used to interpret a high/strong, weak/low or inverse relationship. Correlation between variables is not a cause-and-effect relationship. In essence, two things can be correlated without there being a causal relationship. The results of the correlation can be found in Appendix A. The pink highlighted columns ( ) identified in Appendix A, represent variables that correlated with their respective questions from each row.

Results of Correlation Analysis

This section discusses the results of the correlation analysis that was conducted. The correlation looks at the strength of the relationship between two or more variables. A negative correlation indicates an inverse relationship in which as one variable increases the second variable decreases. The analyses reveal that not all variables had associations with one another. The numbers identified in Appendix A on pg. 44 below corresponds to the highlighted in pink ( ) in Appendix A. The variables with evidence of a relationship or association are identified and discussed below:

1. There is a relationship between total acreage and number of people employed.

The Seaport Planning Activities of American Association of Port Authorities (AAPA)-member Seaports ("Portganisms")

2. There is a relationship between number of people employed at the seaport and the total port acreage. *The more acreage within a port's jurisdictional area the more employees the seaport tends to employ.*
3. There is a relationship between how often a seaport's Master Plan gets updated and if the seaport has an adopted Master plan.
4. There is a relationship between seaports with adopted Master Plans, how often those Master Plans are updated and the preparation of an Affordability Analysis for Master plan projects.
5. There is a relationship between the number of people employed at a seaport and formation of a seaport planning section or planning division.
6. There is a relationship between the primary department where planning occurs and the total number of people the seaport employs. *In essence based on results of #6 below on pg. 43, the analysis show there is an inverse relationship between the total number of people employed at the port, and the primary location where planning-related activities are conducted. In essence, the more employees the seaport employs the less likely there is a primary department where planning occurs.*
7. There is an inverse relationship between the number of planners employed in the seaport and whether the Master Plan is adopted. *Seaports with Master/Vision Plans did not correlate with the number of planners working for that seaport.*
8. There is a direct relationship between the number of seaport planners employed specifically within a planning department and the total number of people employed at the seaport.
9. There is evidence to suggest a relationship between the percentage of Master plan-related work conducted by in-house planners and the number of people employed in the seaport.
10. There is a relationship between the total number of staff employed at the seaport and the number of members serving on the Commission/Board.
11. There is a relationship between number of employees and the primary department where planning occurs.

The Seaport Planning Activities of American Association of Port Authorities (AAPA)-member Seaports ("Portganisms")

12. There is a relationship between the number of people employed and the availability of a planning section or planning division.
13. There is evidence of a relationship between adoption of the Master Plan and whether an Affordability Analysis is conducted for Master Plan projects.
14. There is evidence of relationship between how often a Master plan is adopted and whether an Affordability Analysis is utilized for Master plan projects.
15. There is also evidence suggesting a relationship between a seaport's webpage or website availability dedicated to planning-related activities and Affordability Analysis. In addition a web page dedicated to planning activities also correlates with planning activities occurring under a seaport's planning department or planning division.
16. There is also evidence suggesting a relationship between a webpage or website dedicated to planning-related activities, Affordability Analysis conducted for Master Plan projects, planning activities occurring mostly under Planning Department or Planning Section or Division and the primary Department where planning occurs.
17. There is evidence of relationship on whether planning activities are mostly outsourced and the primary department where planning occurs. The results reveal that planning activities primarily occurring under the Chief Executive/Port Director's Office correlates with the preparation of an Affordability Analysis, creation of a planning section and a planning web page.
18. There is evidence of an inverse relationship between the Port Director's office as the primary department where planning occurs and planning activities under a Planning Department or Section. There is an inverse relationship between planning activities mostly outsourced and whether planning occurred in the Port Director/Chief Executive Office.
19. There is evidence of a relationship between the number of seaport planners employed by a seaport, planning activities occurring under a Planning Department or Division and the primary department where planning occurs. This suggest that an increase in the number of seaport employees is related to the

likelihood that planners hired will be staffed in Planning Department and not within the Port Director/Chief Executive's Offices.
20. Planners employed specifically within a Planning Department or Division correlate with planners conducting planning-related activities within the Port Director/Chief Executive Office.
21. There is a relationship between the total number of people employed by the seaport and the number of Board/Commission members.
22. There is a relationship between planners employed and planning-activities occurring under a Planning Department or Planning Division and the percentage of Master Plan work that is outsourced to a consulting firm.
23. Association exists between the Planning Section and planners specifically employed within a Planning Department and the percentage of Master-Plan work conducted by in-house Planners.
24. There is an association between planners employed at the port within Planning Departments and planners working primarily in a Port Director/Chief Executive's office.
25. There is a relationship between seaport planners and the percentage of Master-Plan related in-house work.
26. There is a relationship between the planners employed in the Chief Executive/Port Director's Office and total number of planners working for the seaport.

Crosstabs and Chi-Square Analysis

A chi-square test was next conducted among the variables in the survey. Several variables were used to test the independence or whether the variables are related or have association. An analysis was conducted to determine if there are relationships between several ordinal or nominal variables or questions in the survey. The chi-square test was conducted to determine the association between a seaport's primary source of financing projects identified in its Master Plan and conduct of an "Affordability Analysis."

The Seaport Planning Activities of American Association of Port Authorities (AAPA)-member Seaports ("Portganisms")

The primary funding sources identified in the survey were federal, state, local/county funds, and Public/Private Partnerships (P3). The chi-square analysis identified some very interesting results among the seaports' use of federal funds as a primary source to finance projects in the Master Plan and whether an Affordability Analysis is conducted. There is no evidence to support a relationship between state and local/county funds as primary sources. However, there is very strong evidence supporting a relationship or association between utilizing of federal funds for projects and the conduct of an Affordability Analysis for Master Plan projects. This relationship was statistically significant at the .058 level (See Appendix B). The chi-square analysis also determined there is evidence to support the relationship between using P3 as primary funding source of financing Master Plan projects and conducting an Affordability Analysis. This relationship was statistically significant at the .010 level (See Appendix B).

Results of Chi-Square Test

1. Ports whose significant funding for Master Plan projects come from federal funds have an effect on whether an Affordability Analysis/Decisions Matrix is conducted for Master Plan projects.

2. There is a relationship between seaports whose primary funding of Master Plan projects are based on Public Private Partnerships (P3) and the conduct of an Affordability Analysis.

3. There is a relationship between seaports that raise capital for port projects through port revenue or floating bonds and stakeholder involvement in the Master Plan process specifically stakeholder involvements from outreach events and forums.

The Seaport Planning Activities of American Association of Port Authorities (AAPA)-member Seaports ("Portganisms")

Implications and Conclusions

This research was concerned with identifying characteristics of port planning activities within AAPA-member seaports, adoption by seaports of a Master Plan, and funding sources for projects identified in the Master and CIP Plans. Seaports are unique and are living "organisms", also known as portganisms *capable of response to stimuli, reproduction, growth and development, and maintenance as a stable whole.* However, in order for a seaport to sustain and develop, it requires planning. Seaport planners play essential roles in helping to facilitate a seaport's development, and growth. Although several seaports reported not having official planning positions are their respective seaports, there are still planning being undertaken at those seaports. The research evidences the relationship between where planning occurs at a seaport and the percentage of work that was outsourced to professional consulting firms. Planning that occurs within a port's Planning Department or Planning Divisions results in less likelihood of planning-related work and, specifically, Master Plan-related work being outsourced. Such implications suggest that primary Master Plan planning activities within non-planning departments tend to be provided by professional consulting firms.

In as much as planners at seaports are essential, the results of the analysis suggest planning activities under the management of Chief Executive/Port Director's offices were most likely to be outsourced. However, the number of people employed at seaport also plays a role in determining the availability of a Planning Department or Division. A few seaports reported that although they had employees whose titles were planners, the adoption of Master Plans were not being conducted at their seaports. The analysis shows that development and adoption of a seaport Master Plan correlated with the total number of people employed at the seaport, existence of a Master Plan, and preparation of an Affordability Analysis. Funding sources also play a major role in Master Plan projects and affordability to pay for those projects, including the availability of federal funding, P3, and stakeholder involvement.

The Seaport Planning Activities of American Association of Port Authorities (AAPA)-member Seaports ("Portganisms")

Significance of the Research to Seaports and AAPA

The significance of this research lies in the quantitative contribution it makes to AAPA. It makes a contribution to future AAPA PPM candidates and to the paucity of evaluation studies on current planning and planning-related activities in seaports. The research also contributes to the current limited pool of analysis and research that is demanded by governmental agencies such as the Maritime Administration, the agency within the U.S. Department of Transportation dealing with waterborne transportation, (MARAD), Florida Department of Transportation (FDOT), and professional associations such as American Planning Association and American Association of Port Authorities (AAPA) on the planning activities and characteristics of planning in U.S. and Canadian seaports. This research also provides information to policy makers, specifically those in the marine industry, about the additional roles which seaports play as economic engines.

REFERENCES

Correlation Studies, Retrieved on December 10, 2013 from ttp://www.psychologyandsociety.com/correlationcausation.html

Moore, Charles (1921). "XXV "Closing in 1911–1912"". *Daniel H. Burnham, Architect, Planner of Cities, Volume 2*. Boston, Massachusetts: Houghton Mifflin. p. 1921.

U.S. Public Port Infrastructure Investment Survey 2012-2016. Retrieved on December 10, 2013 from,http://aapa.files.cms-plus.com/2012%20AAPA%20Port%20Infrastructure%20Spending%20Survey%20Summary.pdf

APPENDIX A: THE CORRELATION TABLE

The Seaport Planning Activities of American Association of Port Authorities (AAPA)-member Seaports ("Portganisms")

Inferential Statistics: Correlation Analysis
Results of Correlation

		Total Acreage	Actual Employees	People employed	Adopted Master Plan	How often is MP updated	CIP or 5-Y Project in MP
Total Acreage 1	Pearson Correlation	1	-.011	.448**	.082	-.087	.021
	Sig. (2-tailed)		.946	.004	.616	.591	.898
	N	40	40	40	40	40	38
Actual or Estimated Employees	Pearson Correlation	-.011	1	-.012	.300	-.289	-.131
	Sig. (2-tailed)	.946		.941	.060	.071	.433
	N	40	40	40	40	40	38
People employed 2	Pearson Correlation	.448**	-.012	1	-.141	.050	-.033
	Sig. (2-tailed)	.004	.941		.384	.758	.843
	N	40	40	40	40	40	38
Adopted Master Plan	Pearson Correlation	.082	.300	-.141	1	-.591**	-.254
	Sig. (2-tailed)	.616	.060	.384		.000	.124
	N	40	40	40	40	40	38
How often is MP updated 3	Pearson Correlation	-.087	-.289	.050	-.591**	1	-.062
	Sig. (2-tailed)	.591	.071	.758	.000		.713
	N	40	40	40	40	40	38
CIP or 5-Y Project in MP	Pearson Correlation	.021	-.131	-.033	-.254	-.062	1
	Sig. (2-tailed)	.898	.433	.843	.124	.713	
	N	38	38	38	38	38	38
Affordability Analysis Used for MP Projects 4	Pearson Correlation	-.004	-.090	.115	-.382*	.339*	.007
	Sig. (2-tailed)	.979	.585	.485	.017	.035	.967
	N	39	39	39	39	39	38
Website/webpge dedicated to the Port MP	Pearson Correlation	-.273	.012	-.032	-.246	.147	.100
	Sig. (2-tailed)	.089	.943	.843	.126	.367	.549

The Seaport Planning Activities of American Association of Port Authorities (AAPA)-member Seaports ("Portganisms")

	N	40	40	40	40	40	38
Planning activities under Planning Department/Committee	Pearson Correlation	-.057	-.105	.261	-.246	.147	-.172
	Sig. (2-tailed)	.728	.520	.104	.126	.367	.302
	N	40	40	40	40	40	38
Planning section or division 5	Pearson Correlation	-.048	-.023	.409**	-.270	.181	-.275
	Sig. (2-tailed)	.770	.889	.009	.092	.263	.095
	N	40	40	40	40	40	38
Planning activities mostly outsourced	Pearson Correlation	.013	.056	-.241	.201	-.037	-.111
	Sig. (2-tailed)	.936	.730	.134	.214	.822	.506
	N	40	40	40	40	40	38
Primary /PD Department where planning occurs 6	Pearson Correlation	.284	-.365*	.322*	-.073	.104	-.166
	Sig. (2-tailed)	.080	.022	.046	.659	.527	.327
	N	39	39	39	39	39	37
Planners employed in the Port 7	Pearson Correlation	.134	-.147	.428**	-.334*	.109	-.112
	Sig. (2-tailed)	.410	.366	.006	.035	.504	.502
	N	40	40	40	40	40	38
Planners employed specifically within the Planning Dpt/Division 8	Pearson Correlation	.112	-.150	.416**	-.261	.117	-.170
	Sig. (2-tailed)	.491	.355	.008	.104	.473	.308
	N	40	40	40	40	40	38
% of MP by in-house staff in 2010-2013 FY 9	Pearson Correlation	.104	.233	.341*	.020	-.176	-.082
	Sig. (2-tailed)	.546	.171	.042	.908	.306	.640
	N	36	36	36	36	36	35
% of MP by consultants in 2010-2013 FY	Pearson Correlation	.141	-.187	-.034	-.067	-.118	.256
	Sig. (2-tailed)	.407	.268	.843	.695	.485	.132
	N	37	37	37	37	37	36
Members serving on the Port Authority/County Board 10	Pearson Correlation	.183	.175	.375*	.024	-.221	.073
	Sig. (2-tailed)	.259	.279	.017	.883	.171	.662
	N	40	40	40	40	40	38

The Seaport Planning Activities of American Association of Port Authorities (AAPA)-member Seaports ("Portganisms")

Correlation Results

		Affordability Analysis Used for MP Projects	Website/ webpage dedicated to the Port MP	Planning activities under Planning Deprtmnt /Committtee	Planning section or division	Planning activities mostly outsourced	Primary Department where planning occurs
Total Acreage	Pearson Correlation	-.004	-.273	-.057	-.048	.013	.284
	Sig. (2-tailed)	.979	.089	.728	.770	.936	.080
	N	39	40	40	40	40	39
Actual Employees 11	Pearson Correlation	-.090	.012	-.105	-.023	.056	-.365*
	Sig. (2-tailed)	.585	.943	.520	.889	.730	.022
	N	39	40	40	40	40	39
People employed 12	Pearson Correlation	.115	-.032	.261	.409**	-.241	.322*
	Sig. (2-tailed)	.485	.843	.104	.009	.134	.046
	N	39	40	40	40	40	39
Adopted Master Plan 13	Pearson Correlation	-.382*	-.246	-.246	-.270	.201	-.073
	Sig. (2-tailed)	.017	.126	.126	.092	.214	.659
	N	39	40	40	40	40	39
How often is MP updated 14	Pearson Correlation	.339*	.147	.147	.181	-.037	.104
	Sig. (2-tailed)	.035	.367	.367	.263	.822	.527
	N	39	40	40	40	40	39
CIP or 5-Y Project in MP	Pearson Correlation	.007	.100	-.172	-.275	-.111	-.166
	Sig. (2-tailed)	.967	.549	.302	.095	.506	.327

The Seaport Planning Activities of American Association of Port Authorities (AAPA)-member Seaports ("Portganisms")

		Affordability Analysis Used for MP Projects	Website/webpge dedicated to the Port MP 15	Planning activities under Planning Dpartmnt/Commi tttee	Planning Section or Division 16	Planning activities mostly outsourced 17	Primary Department where planning occurs/Port Director's office 18
Affordability Analysis Used for MP Projects	Pearson Correlation	1	.335*	.224	.372*	.083	.031
	Sig. (2-tailed)		.037	.171	.020	.614	.852
	N	38	38	38	38	38	37
Website/webpge dedicated to the Port MP 15	Pearson Correlation	.335*	1	.430**	.498**	-.207	.061
	Sig. (2-tailed)	.037		.006	.001	.201	.710
	N	38	39	39	39	39	38
Planning activities under Planning Dpartmnt/Commi tttee	Pearson Correlation	.224	.430**	1	.722**	-.207	.430**
	Sig. (2-tailed)	.171	.006		.000	.201	.006
	N	39	40	40	40	40	39
Planning Section or Division 16	Pearson Correlation	.372*	.498**	.722**	1	-.244	.399*
	Sig. (2-tailed)	.020	.001	.000		.130	.012
	N	39	40	40	40	40	39
Planning activities mostly outsourced 17	Pearson Correlation	.083	-.207	-.207	-.244	1	-.522**
	Sig. (2-tailed)	.614	.201	.201	.130		.001
	N	39	40	40	40	40	39
Primary Department where planning occurs/Port Director's office 18	Pearson Correlation	.031	.061	.430**	.399*	-.522**	1
	Sig. (2-tailed)	.852	.710	.006	.012	.001	
	N	38	39	39	39	39	39
Planners employed in the Port 19	Pearson Correlation	.180	.302	.471**	.651**	-.235	.453**
	Sig. (2-tailed)	.272	.058	.002	.000	.145	.004
	N	39	40	40	40	40	39
Planners	Pearson	.161	.290	.535**	.713**	-.138	.434**

The Seaport Planning Activities of American Association of Port Authorities (AAPA)-member Seaports ("Portganisms")

employed specifically within the Planning Dpt/Division [20]	Correlation Sig. (2-tailed)		.327	.070	.000	.000	.395	.006
	N		39	40	40	40	40	39
% of MP by in-house staff in 2010-2013 FY	Pearson Correlation		-.158	.158	.485**	.419*	-.363*	.303
	Sig. (2-tailed)		.357	.357	.003	.011	.030	.073
	N		36	36	36	36	36	36
% of MP by consultants in 2010-2013 FY	Pearson Correlation		.248	.025	-.248	-.074	.150	-.209
	Sig. (2-tailed)		.140	.886	.140	.665	.377	.221
	N		37	37	37	37	37	36
Members serving on the Port Authority/County Board	Pearson Correlation		-.155	-.150	-.064	-.042	-.270	-.016
	Sig. (2-tailed)		.347	.355	.693	.796	.092	.922
	N		39	40	40	40	40	39

The Seaport Planning Activities of American Association of Port Authorities (AAPA)-member Seaports ("Portganisms")

Correlations

		Planners employed in the Port	Planners employed specifically within the Planning Dpt/Divsion	% of MP by in-house staff in 2010-2013 FY	% of MP by consultants in 2010-2013 FY	Members serving on the Port Authority/ County Board
Total Acreage	Pearson Correlation	.134	.112	.104	.141	.183
	Sig. (2-tailed)	.410	.491	.546	.407	.259
	N	40	40	36	37	40
Actual or Estimated Employees	Pearson Correlation	-.147	-.150	.233	-.187	.175
	Sig. (2-tailed)	.366	.355	.171	.268	.279
	N	40	40	36	37	40
People employed	Pearson Correlation	.428**	.416**	.341*	-.034	.375*
	Sig. (2-tailed)	.006	.008	.042	.843	.017
	N	40	40	36	37	40
Adopted Master Plan	Pearson Correlation	-.334*	-.261	.020	-.067	.024
	Sig. (2-tailed)	.035	.104	.908	.695	.883
	N	40	40	36	37	40
How often is MP updated	Pearson Correlation	.109	.117	-.176	-.118	-.221
	Sig. (2-tailed)	.504	.473	.306	.485	.171
	N	40	40	36	37	40
CIP or 5-Y Project in MP	Pearson Correlation	-.112	-.170	-.082	.256	.073
	Sig. (2-tailed)	.502	.308	.640	.132	.662
	N	38	38	35	36	38
Affordability Analysis Used for MP Projects	Pearson Correlation	.180	.161	-.158	.248	-.155
	Sig. (2-tailed)	.272	.327	.357	.140	.347
	N	39	39	36	37	39
Website/webpage dedicated to the Port MP	Pearson Correlation	.302	.290	.158	.025	-.150
	Sig. (2-tailed)	.058	.070	.357	.886	.355
	N	40	40	36	37	40

The Seaport Planning Activities of American Association of Port Authorities (AAPA)-member Seaports ("Portganisms")

Planning activities under Planning Department/Committee 22	Pearson Correlation	.471**	.535**	.485**	-.248	-.064
	Sig. (2-tailed)	.002	.000	.003	.140	.693
	N	40	40	36	37	40
Planning section or division 23	Pearson Correlation	.651**	.713**	.419*	-.074	-.042
	Sig. (2-tailed)	.000	.000	.011	.665	.796
	N	40	40	36	37	40
Planning activities mostly outsourced	Pearson Correlation	-.235	-.138	-.363*	.150	-.270
	Sig. (2-tailed)	.145	.395	.030	.377	.092
	N	40	40	36	37	40
Primary Department where planning occurs 24	Pearson Correlation	.453**	.434**	.303	-.209	-.016
	Sig. (2-tailed)	.004	.006	.073	.221	.922
	N	39	39	36	36	39
Planners employed in the Port 25	Pearson Correlation	1	.917**	.362*	-.017	.016
	Sig. (2-tailed)		.000	.030	.921	.922
	N	40	40	36	37	40
Planners employed specifically within the Planning Dpt/Division 26	Pearson Correlation	.917**	1	.475**	-.130	-.092
	Sig. (2-tailed)	.000		.003	.443	.572
	N	40	40	36	37	40
% of MP by in-house staff in 2010-2013 FY	Pearson Correlation	.362*	.475**	1	-.542**	.130
	Sig. (2-tailed)	.030	.003		.001	.449
	N	36	36	36	36	36
% of MP by consultants in 2010-2013 FY	Pearson Correlation	-.017	-.130	-.542**	1	-.004
	Sig. (2-tailed)	.921	.443	.001		.983
	N	37	37	36	37	37
Members serving on the Port Authority/County Board	Pearson Correlation	.016	-.092	.130	-.004	1
	Sig. (2-tailed)	.922	.572	.449	.983	
	N	40	40	36	37	40

**. Correlation is significant at the 0.01 level (2-tailed).
*. Correlation is significant at the 0.05 level (2-tailed).

The Seaport Planning Activities of American Association of Port
Authorities (AAPA)-member Seaports ("Portganisms")

APPENDIX B: CHI-SQUARE ANALYSIS

The Seaport Planning Activities of American Association of Port Authorities (AAPA)-member Seaports ("Portganisms")

Appendix B
Inferential Statistics: Chi-Square
Results of Chi-Square

Primary Source 1: Federal * Affordability Analysis Used for MP Projects

Chi-Square Table 1

		Affordability Analysis Used for MP Projects		Total
		No	Yes	
Primary Source 1: Federal	No	15	19	34
	Yes	0	5	5
Total		15	24	39

Chi-Square Tests

	Value	df	Asymp. Sig. (2-sided)	Exact Sig. (2-sided)	Exact Sig. (1-sided)
Pearson Chi-Square	3.585[a]	1	.058		
Continuity Correction[b]	1.963	1	.161		
Likelihood Ratio	5.307	1	.021		
Fisher's Exact Test				.136	.074
Linear-by-Linear Association	3.493	1	.062		
N of Valid Cases	39				

a. 2 cells (50.0%) have expected count less than 5. The minimum expected count is 1.92.
b. Computed only for a 2x2 table

Primary Source 4: P3 (Public Private Partnerships) & Affordability Analysis Used for MP Projects

Chi-Square Table 2

		Affordability Analysis Used for MP Projects		Total
		No	Yes	
Primary Source 4: P3 (Public Private Partnerships)	No	5	18	23
	Yes	10	6	16
Total		15	24	39

The Seaport Planning Activities of American Association of Port Authorities (AAPA)-member Seaports ("Portganisms")

Chi-Square Tests

	Value	df	Asymp. Sig. (2-sided)	Exact Sig. (2-sided)	Exact Sig. (1-sided)
Pearson Chi-Square	6.624ª	1	.010		
Continuity Correction[b]	5.013	1	.025		
Likelihood Ratio	6.715	1	.010		
Fisher's Exact Test				.018	.012
Linear-by-Linear Association	6.454	1	.011		
N of Valid Cases	39				

a. 0 cells (.0%) have expected count less than 5. The minimum expected count is 6.15.
b. Computed only for a 2x2 table

Another chi-square analysis was conducted to determine the effect of the stakeholder's involvement in the Master Plan process and how the port raises capital for improvements or for projects inclusive of Master planning projects.

Stakeholders' Involvement in the MP process- Outreach Events/Forums & Raise capital by Port Revenue / Bonds

Chi-Square Table 3

		Raise capital by Port Revenue / Bonds		Total
		No	Yes	
Stakeholders' Involvement in the MP process- Outreach Events/Forums	No	2	17	19
	Yes	8	11	19
Total		10	28	38

Chi-Square Tests

	Value	df	Asymp. Sig. (2-sided)	Exact Sig. (2-sided)	Exact Sig. (1-sided)
Pearson Chi-Square	4.886ª	1	.027		
Continuity Correction[b]	3.393	1	.065		
Likelihood Ratio	5.151	1	.023		
Fisher's Exact Test				.062	.031
Linear-by-Linear Association	4.757	1	.029		
N of Valid Cases	38				

a. 0 cells (.0%) have expected count less than 5. The minimum expected count is 5.00.
b. Computed only for a 2x2 table

The Seaport Planning Activities of American Association of Port
Authorities (AAPA)-member Seaports ("Portganisms")

APPENDIX C: AAPA SEAPORTS SURVEY INSTRUMENT

The Seaport Planning Activities of American Association of Port Authorities (AAPA)-member Seaports ("Portganisms")

PROFESSIONAL PORT MANAGER® (PPM)
AAPA Member Seaport Survey

This survey is about the current planning activities undertaken at your seaport. The information gathered may help provide information on the characteristics of planning activities at seaports. Your answers will be kept completely anonymous and confidential. Results will be reported in the aggregate. The questions take about **7 minutes** *of your or your planning staff's time. Please place an X in the box for the appropriate responses. Thank you. Your time is very much appreciated*

What is the name of your Seaport?_____

Your Name (Optional)_____

What is your title?_____

1. What is your department?
   _____

2. What is your division?
   _____

3. What is your gender?   Male ☐   Female ☐

4. What is your age?   18-25 ☐   26-35 ☐   36-45 ☐   46-55 ☐   56-65 ☐   65+ ☐

5. How long have you worked for the Port?
   ☐ Less than 2 years
   ☐ 2 to 7 years
   ☐ 8 to 15 years
   ☐ More than 15 years

6. Is your port a landlord port?   Yes ☐   No ☐

The Seaport Planning Activities of American Association of Port Authorities (AAPA)-member Seaports ("Portganisms")

7. Are you an operational port? Yes ☐   No ☐

8. What is the total acreage of your seaport under the Port Authority's Jurisdiction?

    ☐ Less than 1000
    ☐ 1001 to 5000 acres
    ☐ 5001 to 10,000 acres
    ☐ More than 10,000 acres

9. Approximately how many people are employed by your seaport? (Please give estimated figures if actuals are not known).

    ☐ Actual   Estimated ☐

    ☐ Less than 25
    ☐ 25 to 99
    ☐ 100 to 500
    ☐ More than 500

10. Does your seaport have an adopted Master Plan? ☐ Yes   No ☐

11. How often does your seaport update the Master Plan?
    ☐ Every 2 years
    ☐ Every 2 to 5 years
    ☐ Every 5 to 10 years
    ☐ More than 10 years

12. Which department is responsible for managing and updating the Port's Master Plan?
    ☐ Port Director's/Chief Executive Office
    ☐ Public Works
    ☐ Legal Department
    ☐ Environmental Management/Office of Sustainability
    ☐ Port Authority/Port Commissioner's Office
    ☐ Engineering Department
    ☐ Operations Department
    ☐ Other, please list _____

13. Does your seaport Master Plan have a Capital Improvement Plan (CIP) or a 5-Year project component? ☐ Yes   No ☐

The Seaport Planning Activities of American Association of Port Authorities (AAPA)-member Seaports ("Portganisms")

14. How does the port raise capital for improvements or capital projects?

   _____

15. What is the primary source of funding for projects identified within your Port's Master Plan?

   ☐ Federal Funds (e.g., TIGER Grants)
   ☐ State Funds
   ☐ County/Local Funds
   ☐ Public/Private Partnerships

16. Is there an "Affordability Analysis/Decision-Matrix" prepared/conducted when prioritizing which Master Plan projects to implement? ☐ Yes   No ☐

17. Do you have a web page or website dedicated solely to your Port's Master/Vision Plan? ☐ Yes   No ☐

Please tell us about your Seaport Planning Department, Planning Division or Planning Section

18. Are your seaport planning activities conducted under a Planning Department?

   Yes ☐   No ☐

19. Does your Port have a planning section or a planning division? Yes ☐   No ☐

20. Are your seaport planning activities mostly outsourced to a professional planning and consulting firms? Yes ☐   No ☐

21. If no, please check one, which is the primary department where planning activities, such as adoption of the Master Plan, are conducted?

   ☐ Public Works
   ☐ Engineering Department
   ☐ Port Director's Office
   ☐ Port Authority Commission Staff Members
   ☐ Environmental Department/Sustainability Department

22. How many planners are employed in the Port?

The Seaport Planning Activities of American Association of Port Authorities (AAPA)-member Seaports ("Portganisms")

- [ ] 1 to 5
- [ ] 6 to 10
- [ ] 11 to 15
- [ ] More than 15

23. From question 22 above, how many seaport planners are employed specifically within your Port's Planning Department or Planning Division?

- [ ] 1 to 5
- [ ] 6 to 10
- [ ] 11 to 15
- [ ] More than 15

24. Is managing and adopting the Master Plan the primary function of the seaport planning staff? Yes ☐ No ☐

25. Please indicate the sources of funding for your financing plan. Who bears the ensuing costs for projects that are implemented at the seaport?

- [ ] Federal Government
- [ ] Regional Government
- [ ] Local or Municipality
- [ ] International Sector
- [ ] Private Sector
- [ ] Others, please lists _____

26. Do you align the financing of the Master Plan projects with your current budget cycle (as part of the annual Operating budgets)? ☐ Yes No ☐

27. From 2010-2013 fiscal year, what percent of the Board approved Master-plan related projects were primarily implemented or prepared by in-house staff?

- [ ] 0% to 25%
- [ ] 26% to 50%
- [ ] 51% to 75%
- [ ] 76% to 100%

28. From 2010-2013 fiscal year, what percent of Board approved Masterplan-related approved projects were primarily implemented or prepared by a consultant or outsourced firm?

- [ ] 0% to 25%

The Seaport Planning Activities of American Association of Port Authorities (AAPA)-member Seaports ("Portganisms")

☐ 26% to 50%
☐ 51% to 75%
☐ 76% to 100%

29. How many members serve on your Port Authority/County Board?
    ☐ 1 to 5
    ☐ 6 to 10
    ☐ 7 to 11
    ☐ 12 to 15
    ☐ More than 15

30. In terms of outreach, how are stakeholders (experts, citizens, and city/rural/urban dwellers) mostly involved in the Master planning process?
    ☐ Involved in formal meetings
    ☐ Outreach events, forums
    ☐ Consultation with port staff
    ☐ Oral participation
    ☐ Written participation

I want morebooks!

Buy your books fast and straightforward online - at one of the world's fastest growing online book stores! Environmentally sound due to Print-on-Demand technologies.

Buy your books online at

www.get-morebooks.com

Kaufen Sie Ihre Bücher schnell und unkompliziert online – auf einer der am schnellsten wachsenden Buchhandelsplattformen weltweit! Dank Print-On-Demand umwelt- und ressourcenschonend produziert.

Bücher schneller online kaufen

www.morebooks.de

OmniScriptum Marketing DEU GmbH
Heinrich-Böcking-Str. 6-8
D - 66121 Saarbrücken

Telefax: +49 681 93 81 567-9

info@omniscriptum.de
www.omniscriptum.de

www.ingramcontent.com/pod-product-compliance
Lightning Source LLC
Chambersburg PA
CBHW031545210526
45464CB00003B/1161